BEI GRIN MACHT SICH IHR WISSEN BEZAHLT

- Wir veröffentlichen Ihre Hausarbeit,
  Bachelor- und Masterarbeit

- Ihr eigenes eBook und Buch -
  weltweit in allen wichtigen Shops

- Verdienen Sie an jedem Verkauf

Jetzt bei www.GRIN.com hochladen und kostenlos publizieren

Bruno Yote

Regional Competition as a „relational problem?

The Example of the Aerospace Industry in Toulouse, Seattle and north-west England

GRIN Verlag

Bibliografische Information der Deutschen Nationalbibliothek:

Die Deutsche Bibliothek verzeichnet diese Publikation in der Deutschen National-
bibliografie; detaillierte bibliografische Daten sind im Internet über http://dnb.d-
nb.de/ abrufbar.

Impressum:

Copyright © 2009 GRIN Verlag GmbH
Druck und Bindung: Books on Demand GmbH, Norderstedt Germany
ISBN: 978-3-656-02467-5

Dieses Buch bei GRIN:

http://www.grin.com/de/e-book/179907/regional-competition-as-a-relational-pro-
blem

GRIN - Your knowledge has value

Der GRIN Verlag publiziert seit 1998 wissenschaftliche Arbeiten von Studenten, Hochschullehrern und anderen Akademikern als eBook und gedrucktes Buch. Die Verlagswebsite www.grin.com ist die ideale Plattform zur Veröffentlichung von Hausarbeiten, Abschlussarbeiten, wissenschaftlichen Aufsätzen, Dissertationen und Fachbüchern.

Besuchen Sie uns im Internet:

http://www.grin.com/

http://www.facebook.com/grincom

http://www.twitter.com/grin_com

Bruno Yote (2009)

Regional Competition as a 'relational' problem? – The Example of the Aerospace Industry in Toulouse, Seattle and north-west England

Table of contents

Table of figures

1. Introduction: An approach to relational geometries

The relational approach goes beyond a mere geographical point of view by emphasizing its focus on the varying forms of relations (e.g. social, cultural) among actors and structures that effect dynamic changes in the spatial organization of economic activities. Its research topics are economic innovations, cross-company forms of organization and processes of collective-institutional learning.[1]

Hence, core elements of the relational approach are organization (e.g. cluster), evolution (e.g. historical structures), innovation (e.g. technological development) and interaction (e.g. learning, mutual trust) in and between structures and actors.[2]

Following the relational approach, a research of the development of the aerospace industry is of particular interest, due to its concentration as "an assembly and high-technology industry that inevitably involves a high level of inter-company collaboration"[3], its internationalized character and its different development in various regions, which has been significantly influenced by organizational, structural and innovatory changes.

The study reviews those changes by emphasizing on the evolutionary development of the aerospace industry in Toulouse, Seattle and North-west England in terms of historical achievements and internal and external changes.

By linking relational perceptions and empirical results, the study aims to clarify if regional competition in the aerospace industry can be seen as a 'relational' problem.

Therefore, I will first give a brief proposal to amplify the relational approach, based on the work of Yeung (2005). Thereafter, I will present the empirical findings concerning aerospace-related research institutes, processes of internationalization and specification, and the creation of regional and inter-regional networks. This part is based mainly on the works of Hickie (2006) and Niosi/Zhegu (2005), as well as on Internet presentations of the different aerospace-related alliances, firms and locations of the three regions.

The results of the study are going to be presented in a brief conclusion that shows how the ideas of the relational approach can be conceived concerning the aerospace industry of our three regions.

---

[1] cf. Haas/Neumair 2007, pp. 31 f.
[2] cf. Bathelt/Glückler 2003, pp. 36 ff.
[3] Hickie 2006, p. 697.

2. "Rethinking relational economic geography"

In the essay "Rethinking relational economic geography" (2005) Yeung amplifies the above described relational approach by highlighting the importance of networks and network relations: success and prosperity of firms are explained in terms of how inter-firm networks perform "*in relation* to competing networks in the same region and elsewhere (e.g. global competition) ... and the importance of this network in relation to the firms overall transactional activities that often go beyond localized networks"[4]

Furthermore he draws our attention to the importance of power[5] relations and actor-specific practices in the making of economic space and its power geometries: "[m]ediated and realized through actor-specific practice, the emergent power ... provides a major force to drive association and interconnections and to produce socio-spatial outcomes."[6]

Placing the analytical emphasis on heterogeneous configurations of power relations within particular regions, the relational approach conceives the region as a relational construct through which heterogeneous flows of actors, assets and structures coalesce and take place.[7] It analytically focuses on the inherent tension in producing regional development outcomes, and analysis in particular the *relational complementarity* and *relational specificity* of these actors (local and non-local), assets (tangible and intangible) and structures (formal and informal), and their interactive power relations.[8]

This methodological specification allows for an analysis of why some actors (e.g. firms and unions) are more tied to specific regions and therefore likely to contribute to regional development and helps to identify the relational advantage of regions when a particular set of heterogeneous relations might be more beneficial to one region than to another one.[9]

3. The aerospace potential of Toulouse, Seattle and North-west England: a introductory regional presentation

Toulouse is the capital of the Midi-Pyrenees region in south-west France. Its the home of Airbus Industry, a European consortium founded in 1970 between Aerospatiale

---

[4] Yeung 2005, p. 43.
[5] Yeung defines power as "the emergent effects of *social practice among actors* who have the capacity and resources to influence", ibid., p. 44.
[6] ibid., p. 46.
[7] cf. ibid., pp. 47 f.
[8] cf. ibid., pp. 46 ff.
[9] cf. ibid., p. 48.

(France) and Deutsche Airbus (Germany), soon CASA (Spain) and British Aerospace (GB) joined them (1971, 1979). Today Airbus is the greatest employer of the city.[10] Apart from aerospace industry, Toulouse is a centre for electronics, information technology, and biotechnology. It is also the third major French university city with around 110.000 students. The city's growth has been significantly due to the aerospace industry and aerospace related industries and companies.[11]

Figure 1: Toulouse and 'Aerospace Valley'

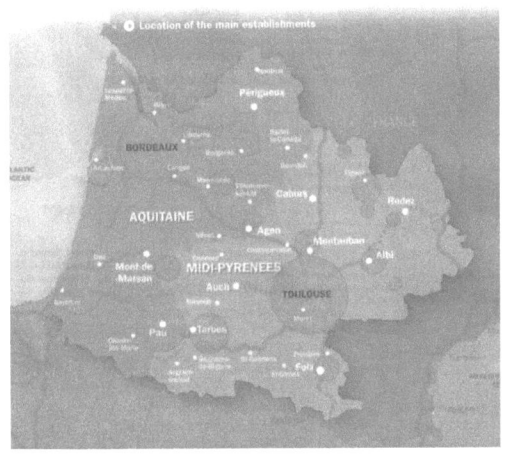

Source: Aerospace Valley (2006), p. 3.

Seattle, in Washington State in the north-west of the USA, has been the home of the Boeing company from as early as 1917 until 2001, than it moved to Chicago, Illinois.[12] However, Seattle is still the headquarters to a number of Boeing companies, most significantly Boeing Commercial Aircraft, and its R&D (research and development) subsidiary Phantom Works, so that the aerospace industry is still highly represented in Seattle. Nevertheless, nowadays the city is also known as the headquarters of Microsoft. Like Toulouse, Seattle is a university city with six universities in or near to it.[13] In Washington State there are about 156 aerospace companies, from the indigenous and basic (e.g. Aeroform, a fabrications company), to subsidiaries of major companies supplying both Boeing (e.g. Saint-Gobain Performance Plastics) and Airbus (e.g. Matsushita Avionics Systems).[14]

---

[10] cf. Niosi/Zhegu 2005, p. 17; cf. Aschenbroich 2006, pp. 50 f.
[11] cf. Hickie 2006, pp. 697 f.
[12] cf. www.boeing.com 2001.
[13]cf. Hickie 2006, p. 698.
[14] cf. ibid.

Figure 2: Washington State

Source: www.holiday-rentals.co.uk (2009)

The *north-west of England* is a region with several aerospace centres and about 600 to 800 aerospace related companies that are prime- and sub-contractors in the aerospace and high-technology engineering industries.[15] Its heart is in central *Lancashire*, around *Preston*, where BAE Systems, the successor company of British Aerospace, has plants at *Warton* and *Samlesbury*, and where Rolls Royce has a plant nearby at *Barnoldswick*. Until recently BAE Systems manufactured regional jets near *Manchester* (in *Chadderton* and *Woodford*). Airbus UK (partly owned by BAE Systems) has a plant at *Broughton* in *North Wales*, close to *Chester*.[16]

Figure 3: North-west England

Source: Greenhalgh (2008), p. 2

---

[15] The numbers differ from over 600 (cf. www.lancashire.gov.uk 2008) to over 800 aerospace companies in North-west England (cf. Hickie 2006, p. 698).

[16] cf. Hickie 2006, p. 698; cf. www.lancashire.gov.uk 2008.

4. The relational approach: Evolution, organization, innovation and interaction

In the first 20 years of the 20th century aerospace industry developed almost simultaneously in Europe and America.[17]

Its pioneers, A.V. Roe, Bill Boeing and Emile Dewoitine (the founder of the Latecoere company), entered the industry on a tiny scale and were competent and fortunate enough to receive an early boost from government contracts. Due to their success their companies "acted as a catalyst, both developing industrially relevant knowledge and skills within the region and attracting them to it."[18]

Between the two World Wars government "dispersal policy" (to avoid enemy bombing) had a significant impact on the geographical and organizational configuration of the aviation industry (especially in Europe), so that Toulouse and the north-west of England became more attractive for investments in the industry because of their distance to the German borders.[19]

For example, the dispersal policy was highly significant for the evolution of North-west England because of the structural implications it had on the region, e.g. in 1938 English Electric re-entered the industry after twelve years of absence in manufacturing aircraft, because the government wished to draw upon the company's capability as a manufacturer upon the local knowledge base, using local supplies of highly skilled industrial labour.[20] Its successfully move into the design of military jets, and the development of new designs and expertise associated to it, formed the basis of 60 years of continuous development at the company's plants around Preston, that today are the heart of BAE Systems' military aviation activities. During the 1940s jet engine development was transferred to the area around Barnoldswick. Nowadays it's the world-leading centre of jet engine development and production under the leadership of Rolls Royce.[21]

Beyond that, two major mergers, both directly as a consequence of government policy (1950s, 1977), leaded to the creation of British Aerospace, that later merged into BAE Systems (1999).[22]

Despite that "it can be argued, that it was the aviation facilities introduced as a result of

---

[17] 1910 in north-west England (A.V. Roe), 1916 in Seattle (Boeing) and 1921 in Toulouse (Latecoere), cf. Hickie 2006, p. 699.
[18] ibid.
[19] cf. ibid.
[20] e.g. from the textile machinery industry, cf. ibid.
[21] currently the turbine blades for the Airbus A380 are manufactured there; cf. ibid., p. 701.
[22] cf. ibid., p. 702.

dispersal policy... that have given the north-west of England both its particular specialism and its long-term global competitiveness in aerospace, enabling the region to preserve and develop its expertise over seven decades."[23]

Nonetheless, after World War II Boeing captured a pre-dominant role in the commercial aircraft industry because government assistance, especially timely military orders, helped Boeing to ride out a number of economic difficulties. However, it has been organizationally quite stable, e.g. tending to take over competitors,[24] and is still the world's largest producer of both military and civil aircraft, hence the largest aerospace company in the world. Apart from Seattle, Long Beach, in California, is the main production site of Boeing. Whereas only the production lines in Seattle represented 75% of its commercial airplane division in 2001.[25]

For Toulouse the aerospace industry was of even greater organizational upheaval between 1936 and 1970. 1936 Dewotoine's company was nationalized as part of SNCA du Midi, that in 1942 merged with Sud Est and in 1958 became part of Sud Aviation. Latter merged into Aerospatiale in 1970 that became part of Airbus Industry (see above).[26] All these mergers form part of a vast learning process in the post-war periods that established the region in the international market and enabled the contribution of key elements of Aerospatiale's expertise to Airbus Industry.[27]

Aerospace as a high-technology industry not only faces technological risks and difficulties, it has also been subject to radical fluctuations in demand and major organizational restructurings, often externally imposed by governments, as the examples show. The current success of the three regions is based upon a high degree of continuity of aircraft development and production activities (of about nearly one century), that are both critical in aerospace for the knowledge development and financial status of the companies. Capital growth has been accompanied by a parallel growth in scientific, technological, commercial and managerial knowledge and skills. The development and enhancement of this knowledge base has both been a response to market forces and a consequence of public policy, intending to preserve and foster the local aerospace industries.[28]

---

[23] ibid., pp. 701 f.
[24] e.g. McDonnell Douglas in 1997, cf. ibid., p. 702.
[25] cf. Niosi/Zhegu 2005, p. 18; cf. Hickie 2006, pp. 702 f.
[26] cf. Hickie 2006, p. 702.
[27] e.g. organizational, marketing and technological lessons learned during the development of the Caravelle (1950s) and the Concorde (1960s), cf. ibid., p. 700.
[28] ibid., pp. 700 ff.

The following chapters should give an overview of significant organizational and structural aspects that propelled the aerospace industry of the three regions and had impacts on socio-spatial outcomes.

4.1 The aerospace clusters of Toulouse, Seattle and north-west England – Research and interaction?

Governments support was essential to the development of the regions', both in developing a scientific and technological knowledge-base of the aerospace industry and in ensuring its supply of highly skilled researchers and engineers. Such support can be given in a variety of ways, for example military production contracts or infrastructural support, e.g. government-funded research institutes.[29]

Research institutes are essential for innovatory steps in the industry's development and form a structural key part of the companies' networks.

The Boeing company has a long history of mutually supportive relationship with the University of Washington at Seattle. Since Boeing's beginnings in 1917, cooperation with the University of Washington was established, and already by 1926, only one of the company's engineers was not a University of Washington graduate. Boeing, in turn, had a vital interest in developing an effective R&D stock and invested *vice versa* in the technological and structural support of the University.[30] The University's Aeronautics and Astronautics Department funded by Washington State and by NASA opened as early as in 1929 and provided Boeing with research-based knowledge, skilled labour, and test facilities.[31] However, Boeing has no support within the state from locally-based NASA research institutes, therefore "[i]t may be argued that Boeing's capacity to prosper technologically, without an industry-focused research centre nearby, may reflect a number of factors, such as its scale and technological ambition, which enabled the company to develop substantial R&D and engineering facilities ... of its own."[32]

The case of Toulouse is quite different. Toulouse was not locally supported by specialist government research and educational institutions until 1969 when ONERA, the national office for aerospace studies and research, set up a research centre (ONERA-

---

[29] cf. ibid., p. 706.
[30] e.g. construction respectively modernization of the University's wind tunnel (1917 & 1936) or upgrade of computer facilities (1980s); cf. ibid., pp. 705 f.
[31] cf. Lee *et al* 2003, pp. 4 f.
[32] Hickie 2006, p. 707.

CERT) in Toulouse "to conduct applied research relevant to the industry",[33] and the *Supaéro grande ecole,* an aeronautics engineering school, was transferred to the city "to provide graduates of a very high calibre to the industry."[34]

Today, Toulouse forms part of the 2005 created Aerospace Valley (see Figure 1), which connects the aerospace industries of the French Midi-Pyrenees and Aquitaine regions. Aerospace Valley concentrates 45% of the total French R&D potential in the sectors of aeronautics, space and embedded systems. In total over 80 specialized public research centres, six universities and twelve *grande ecoles* belong to Aerospace Valley and give jobs to more than 8.500 researchers in aerospace related research programmes. An Aerospace Campus in Toulouse, "which will group on the same site the main education, research and industry participants, with over 1.000 researchers"[35] is under construction. Therefore Aerospace Valley can be considered as a huge innovation and research think tank, rather than a primarily business oriented cluster.[36]

For historically reasons (see above) the main expertise of the region of North-west England is in airframe design and manufacture, rather than in R&D; a major government aerospace research centre is yet still missing. The region's research qualities and facilities developed through engineering programmes of local universities (e.g. Manchester, Liverpool, Bolton, etc.) '"have been largely untapped by the aerospace industry'"[37], especially within the SME (small- and medium-sized enterprises) community; and in fact innovation within the SME's is limited. The "Aerospace Cluster Strategy 2007 to 2017" quotes as a reason for this "that the outputs required and measured by universities from research and technology ... are different to the outputs required by the industry."[38]

However, there are currently steps taken to promote sector-related R&D centres in the region, like for example the creation of an Aerospace Innovation Centre (AIC), a key strategic project of the Northwest Aerospace Alliance (NWAA), that is intended "to provide access to technology and innovation across the supply chain, combining the talents of the region's eight key universities and industry ... in a permanent facility to support companies in aerospace and related sectors."[39]

---

[33] ibid.
[34] ibid.
[35] Aerospace Valley (2006), p. 7.
[36] cf. www.eacp-aero.eu 2009; cf. www.aerospace-valley.com/research 2009.
[37] NWDA 2004, cited in Hickie 2006, p. 707.
[38] Northwest Aerospace Alliance 2007, p. 6.
[39] www.lancashire.gov.uk 2008; cf. Northwest Aerospace Alliance 2007, p. 6.

In summary, the Toulouse region does not only have its own source of highly skilled labour and research-based knowledge, but also access to the national and international research networks. Whereas Boeing seems to be more restricted to the locally-based research institutes and in North-west England the research capacities are yet rudimentary or not customized to the companies' requirements.

4.1.1 Internationalization, specification and continuity of relationships

Since the 1960s the aerospace industry has undergone a structural transformation, due to the advancing technological progress. Production processes have also become more work intensive, hence more expensive. These changes have especially proved hardly to sustain single-handed by the European aerospace companies. The Europeans lacked the technological and financial capabilities of the American aircraft industry, so they had simply no other choice than propelling processes of internationalization and specification, not only because the single costs to each partner are lower due to the share of labour, each company also benefits from the technological know-how, experience and tacit knowledge of its partners. Whereas managing such projects proves to be quite difficult, e.g. agreeing specifications, agreeing work shares, controlling costs.[40]

Organizationally this has meant closing less competitive facilities, and focusing the remaining ones on a narrower range of activities. This may or may not mean a loss of knowledge for the region concerned because in aerospace, knowledge spillovers are technology based and focused on supply chain management linking the major companies and their suppliers. For example the closure of British Aerospace's plant in Preston in the 1990s largely meant a transfer of work and expertise to other neighboring sites, whereas the ending of regional jet manufacturing in Manchester lead to atrophy of relevant knowledge and expertise. As a result the Manchester region will not be involved significantly in the development of the next generation of regional jets.[41] This example shows that regional competence and the regional stock of knowledge need to be sustained and renewed continuously

Major aerospace projects are results of strategic partnerships and international alliances. Airbus is the most complete example therefore because its partners have operated together since 1969/ 1970. One of the milestones in the development of Airbus

---

[40] cf. Niosi/Zhegu 2005, pp. 8 ff.; cf. Hickie 2006, pp. 708 f.
[41] cf. Niosi/Zhegu 2005, pp. 8 f.; cf. Hickie 2006, p. 708.

was the Anglo-French Concorde project between Sud Aviation (and its subsidiary SNECMA) and British Aircraft Corporation (and its subsidiary Rolls Royce). The transfer of know-how was especially beneficial to Sud Aviation and gave the company the possibility to make technological innovations (e.g. in aerodynamics or in metallurgy), but also taught the company critical lessons about the effective structuring and management of international aerospace projects.[42]

Sud Aviation's organizational problems with the Concorde played a pivotal role in the creation of Airbus Industry. The creation itself represented a recognition by European governments and aerospace companies that they needed to coalesce not only their financial but their intellectual resources to be able to compete in the worlds market. Nowadays, after almost 4 decades of working together, the partners have shared common experiences and developed common values, norms of behaviour and tacit understandings, briefly said: a common developing culture. This mutual confidence and shared understanding was critical to the creation of a single company with a unified decision-making structure and for the integration of the partner-companies activities to the point of having a single final assembly line in Toulouse.[43]

4.1.2 Regional networks and the continuity-factor

The increasing specialization and internationalization of the aerospace industry created niche markets, that are on a large scale occupied by specialist subcontractors. A large, complex and sophisticated network of subcontractors is fundamental to the competitiveness of aerospace prime contractors. The lack of strong supplier networks was a significant weakness for the French aviation in the 1950s, whereas Boeing at the same time "has been reckoned to be the largest supporter of small business in the state [Washington, A/N]."[44] Another core capacity of Boeing was that a vast part of its suppliers have been companies of former Boeing employees that set up a specialist business to meet the demands of Boeing. Furthermore, it is important to quote that although American aircraft industry had no involvement in the invention of jet engine, or in the design of earliest jet aircraft, companies like Boeing had well managed, well educated and well resourced design teams, a large body of useful complementary technical knowledge, and an innovatory culture (see above). As a result they were more

---

[42] cf. Schmidt 1997, pp. 142 ff.; cf. Hickie 2006, pp. 700 & 709 f.
[43] cf. Schmidt 1997, pp. 142 ff.; cf. Hickie 2006, pp. 709 f.
[44] Hickie 2006, p. 710.

quickly and successfully able to exploit new scientific and technological knowledge.[45] This capacities, that disembogued in a both innovatory and development culture on a local scale produced an atmosphere of continuity and specificity and have had a huge impact on the consolidation of Boeing as the market's world leader, as well as on spatial outcomes; Seattle became a centre for high-technology products and skills.[46]

A further example is the Latecoere company that has been based in and around Toulouse for over 80 years. Although its function in the regional aviation industry has changed fundamentally, it has developed a number of specialist market niches that were originally developed to meet the needs of the local aerospace *technopole* but are now forming part of the global business. Nevertheless, Latecoere is still the major provider for equipment, subassemblies and consultancy services to the aerospace industry of Toulouse.[47]

Continuity has also been given in North-west England where small, highly specialized engineering companies developed their expertise in cooperation with major companies like BAE Systems (and its predecessor companies) and Rolls Royce.[48] The Goodrich Corporation's plant at Huyton, near Liverpool, supplies components for various aviation control systems since 1982 both to regionally-based companies (e.g. Airbus UK and Rolls Royce) and such as Bombardier (Montreal and Toronto, Canada) and Boeing.[49]

However, the three major aerospace companies (BAE Systems, Airbus UK and Rolls Royce) see themselves confronted with supply difficulties due to "a low grade infrastructure ... [that] limits growth and capability development."[50] To overcome these problems they have worked out jointly with the NWAA the "Aerospace Supply Chain Excellence (ASCE) Programme", which will cost about 8.4 million pounds and last for four years (2006 - 2010). The intention is to develop "a World-Competitive Supply Chain in the North West of England ... [that] will be continually adapted to meet the demands of new products emerging such as the Airbus A380... as well as the needs to transition to a knowledge based economy."[51]

---

[45] Recently it has been argued, that Boeing has demonstrated a risk aversion and a commitment to short-term shareholder value, making it less technologically innovatory, and allowed Airbus to take market leadership with superior products. cf. Hickie 2006, p. 705.
[46] cf. Niosi/Zhegu 2005, pp. 10 f.; cf. Hickie 2006, p. 711.
[47] cf. Hickie 2006, p. 711.
[48] cf. ibid; cf. www.lancashire.gov.uk 2008.
[49] cf. Hickie 2006, p. 711.
[50] Northwest Aerospace Alliance 2007, p. 9.
[51] Ibid., pp. 19 ff.

Nevertheless, BAE Systems, Airbus and Rolls Royce are reckoned to buy-in three-quarters of their requirements in the region and are supported by local players, such as Ferranti Technologies, Aircelle and Ultra Electronics. Yet it appears that the region's potential is valued in different terms by major companies and regional politics: contradictory to the quotation above, for example, on the Homepage of Lancashire it is pointed out that in the north-west the "[local primes find] a whole infrastructure of [SME's] supplying sub-contract products and services to [them] ... Such local capabilities are unsurpassed by any other region in the United Kingdom and effectively represents a world-class centre for excellence and a major regional ... resource."[52]

Despite of all that, it can be argued that the more an aerospace region's prime contractor is embedded in a network of local suppliers, the less likely it is to relocate core activities elsewhere.

4.1.3 International network-relations of Airbus and Boeing

By the end of the 1980s international cooperation between different members of the aircraft related industry became commonplace both for American and European firms. Where inter-organizational relationships are developed skill- and knowledge-flows are transmitted vertically up and down the supply chain (both from supplier to customer and from customer to supplier). But subcontractors may also supply one another with complementary products, developing their own networking relationships, and provide mutual assistance and support.[53]

Today Airbus has more than 150 sites throughout the world, and maintains 16 development and manufacturing facilities in France, Germany, Great Britain and Spain; furthermore it's the world's largest producer of commercial aircraft. The company utilizes a network of some 1500 subcontractors in 30 countries, and spent 14,1 billion Euros on its procurement in 2001. Airbus' largest single provider is the USA with over 800 suppliers located in 40 states.[54]

In the meantime, Toulouse has become a major aerospace cluster with about 500 firms directly linked to aerospace[55] and has attracted other aerospace producers not necessarily linked with civil aircraft, such as Matra and Alcatel (satellite

---

[52] www.lancashire.gov.uk 2008.

[53] cf. Niosi/Zhegu 2005, p 10; cf. Hickie 2006, pp. 710 f.

[54] cf. www.fdimagazine.com; cf. Niosi/Zhegu 2005, p. 17.

[55] e.g. ATR (Franco-Italian manufacturer of turboprops), Messier-Dowty (landing gear for 30 airframes, including Airbus) and Turbomeca (turbines).

telecommunications).[56]

Boeing is somewhat different from other prime contractors, so far as, for decades, it internalized most of its main structural parts.[57] However, engines from its six different models (727 to 777 series) come from all over the world: GE and P&W in the USA, Rolls Royce in the UK and CFM-SNECMA in France. Avionics are supplied most often by Honeywell (USA) and BAE Systems (UK).[58]

Despite of all it is not really arguable that Boeing has produced less international (as well as inter-regional and regional) knowledge transfers, respectively spillovers, than other major aerospace producers, due to its internalized network. But the impacts of the last aircraft crisis in 2001 forced Boeing to accelerate its vertical disintegration and look for foreign partners and locations in order to increase market penetration, as well as to reduce design and production costs. Of the over 400 suppliers of Boeing only a handful have facilities in and around Seattle, most major parts of its aircrafts are acquired from other regions of the USA, and increasingly often, from abroad (especially Asia).[59]

For example, the newest model of the Boeing family, the 7E7, will be sourcing parts from subcontractors worldwide. Therefore the company has joined with more than 20 international systems suppliers to develop technologies and design concepts for the 7E7, among others, Messier-Bugatti (France), Diehl and Liebherr-Aerospace Lindenberg (Germany) and BAE Systems (UK). [60]

Nonetheless, it has recently been argued strongly that the Boeing's increased reliance on major overseas subcontractors has significantly weakened its independent knowledge base, and helped potential rivals to develop their expertise due to the mechanisms of international transfer of technology and knowledge. [61]

The main difference between Toulouse and Seattle is the fact that the regional stock of knowledge and expertise in Toulouse is developed within the cluster and stays within the local Airbus family and its subsidiaries, while for Seattle the outsourcing of key technologies and tacit knowledge could lead to an atrophy of the local stock of knowledge, (as happened to Manchester's jet engine industry, see above).

In the relationships between Toulouse's aerospace industries partners, the (tacit element of) knowledge can also be more easily transmitted without distortion in local

---

[56] cf. Niosi/Zhegu 2005, p. 17.
[57] e.g. its two main production sites (Seattle, Long Beach) Boeing employees represent over 80%.
[58] cf. Niosi/Zhegu 2005, p. 18.
[59] cf. ibid., pp. 18 f.
[60] cf. www.fdimagazine.com 2004.

communication. The risk of "loosing" parts of this knowledge is therefore significantly reduced.

The examples of Toulouse, Seattle and the north-west of England show that stronger relationships and inter-firm cooperation between all aerospace related structures, regions and actors are essential to find the right answers to innovatory steps that have been taken and are going to be taken within the development of new aircrafts, such as the Airbus A380 or Boeing's 7E7 model.

As customers seek closer relationships with suppliers, and by implication become more dependent to them, they tend to narrow their supplier base to fewer companies and have strong interests in developing the expertise of their suppliers, so reinforcing the regional stock of knowledge. Local governments and industry organizations also have vital interests in the fostering of aerospace related skills, the clustering of national aerospace educational and research institutions around Toulouse acts in this way, as well as the NWAA's ASCE-programme.

Figure 4: Relationality and competitiveness in aerospace clusters

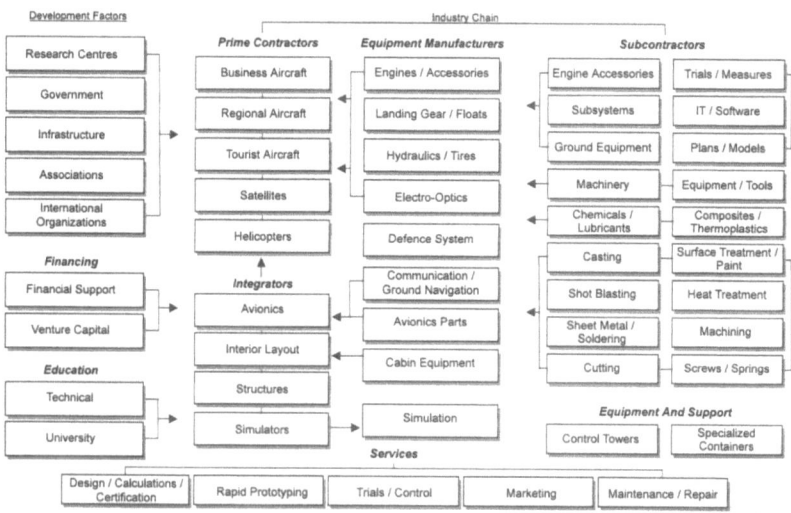

Source: Communauté métropolitaine de Montréal (2004), p.11

---

[61] cf. Niosi/Zhegu 2005, p. 19; cf. Hickie 2006, p. 712.

5. Conclusion: Regional Competition as a 'relational' problem?

How can we then bridge the results of our study concerning the competitive potentials of the three regions and the theoretical ideas of the relational approach?

First, we have seen that there have been significant organizational changes, often influenced by government politics, which favoured the *clustering* of national aircraft industries in particular regions (see p. 3 ff.). Second, innovations in the aerospace industry are highly dependent on *research related* institutes and programmes. The close relationship between Boeing in Seattle and the University of Washington has been proved a crucial aspect to claim the leading position the company has nowadays (see p. 5 f.). Nevertheless, both its restriction to the local stock of knowledge and processes of *internationalization and specification* from the 1960s on changed regional aerospace industries significantly. The beginning *collaborations* between the European aircraft industries culminated in the creation of Airbus that rivaled Boeing's pre-dominant position right from its beginnings. The *emergent power* in the Anglo-French relations drove *associations and interconnections* between and among different actors (firms) and structures (governments, research institutes) and created socio-spatial outcomes of highly *relational specificity and complementarity*: On the one Hand we have Toulouse, embedded in Aerospace Valley's networks, which nowadays is "a huge innovation and research think tank" (see p. 6), on the other hand there is the extremely specialized manufacturing site of North-west England, among others, a world's leader in jet engines (see p. 4 ff.). The examples of Toulouse and North-west England also show us the *inherent tensions* in and between these regions, as regional firms are both complementary and competitive to each other, due to their *network-connections*. Furthermore we have seen that global networks may be beneficial (for example the "common developing culture"/ mutual learning of Airbus connected firms, see p. 8), but can also be a risk for the regional development of aerospace industries as the recent discussion about Boeing's reliance on overseas subcontractors shows (see p. 11).

Hence, is regional competition a relational problem? There seems to be no clear answer to the question. For Toulouse and North-west England regional competition is rather a relational advantage than a problem. The case of Seattle appears to be more differentiated. Regional competition in and between local networks has been beneficial both to Boeing and the region's infrastructure, while Boeings recent intend to be more strongly linked to global network-relations proved to be a little bit "bumpy" as we have seen. Despite all, Boeing is still the market leader and a dominant actor in aerospace business.

Therefore, we can draw the conclusion that the success, hence competitiveness, of aerospace-related actors is not based on the scalar extend of their network-relations, but on the existence of a complex and sophisticated network of long-term relationships having an evolutionary nature where collaboration and competition exist simultaneously.

In summary our results point to the tendency that in a globalized world with global (aerospace-) markets 'the region' can no longer be seen and studied as an 'isolated island' but has to be conceived as a "relational construct through which heterogeneous flows of actors, assets and structures coalesce and take place."[62]

---

[62] Yeung 2005, p. 47.

Bibliography

Monographics:
- Aschenbroich, U. (2006): Markterschließungskonzepte von Airbus und Boeing, Grundlagen und Vergleich. Saarbrücken: VDM Verlag Dr. Müller
- Bathelt, H.; Glückler, J. (2003): Wirtschaftsgeographie: ökonomische Beziehungen in räumlicher Perspektive. Darmstadt: WBG.
- Haas, H.D.; Neumair, S.-M. (2007): Wirtschaftsgeographie. 2. Auflage. Stuttgart: Ulmer.
- Schmidt, A. (1997): Flugzeughersteller zwischen globalem Wettbewerb und internationaler Kooperation, Der Einfluß von Organisationsstrukturen auf die Wettbewerbsfähigkeit von Hochtechnologie-Unternehmen. Berlin: rainer bohn verlag

Essays:
- Hickie, D. (2006): Knowledge and Competitiveness in the Aerospace Industry: The Cases of Toulouse, Seattle and North-west England. In: European Planning Studies, H. 5, pp. 697-716.
- Niosi, N.; Zhegu, M. (2005): Aerospace Clusters: Local or Global Knowledge Spillovers? Available at: http://www.er.uqam.ca/nobel/r21010/document/niosizhegu.pdf [Date: 6/17/09]
- Lee, J.; Eberhardt, D.; Briedenthal, R.; Brucher, A. (2003): A History of the University of Washington Department of Aeronautics and Astronautics 1917 – 2003. Available at: http://www.aa.washington.edu/about/history/AA_History.pdf [Date: 6/17/09]
- Yeung, H.W. (2005): Rethinking relational economic geography. In: Transactions of the Institute of British Geographers, Vol. 30, No. 1, pp. 37-51

Internet sources:
- Aerospace Valley (2006): Aerospace Valley, Midi Pyrenees and Aquitaine. Available at: http://www.aerospace-valley.com/en/brochure.html [Date: 6/17/09]
- Aerospace Valley (2009): Research. Available at: http://www.aerospace-

valley.com/en/the-cluster/research.html [Date: 6/17/09]

- Boeing (2001): Boeing Chooses Chicago as Center of New Corporate
  Architecture. Available at:
  http://www.boeing.com/news/releases/2001/q2/news_release_010510a.html
  [Date: 6/22/09]

- Communauté métropolitaine de Montréal (2004), Aerospace Cluster. Available
  at: http://www.cmm.qc.ca/fileadmin/user_upload/documents/gm_aerospace.pdf
  [Date: 6/11/09]

- European Aerospace Cluster Partnership (2009): Aerospace Valley. Available
  at: http://www.eacp-aero.eu/index.php?id=26 [Date: 6/17/09]

- FDI Magazine (2004): Aerospace: Airbus and Boeing's FDI strategy. Available
  at:
  http://www.fdimagazine.com/news/fullstory.php/aid/709/Aerospace:_Airbus_an
  d_Boeing_92s_FDI_strategy.html [Date: 6/17/09]

- Greenhalgh, B. (2008): NWDA, Cluster Manager, Cluster Development.
  Available at:
  http://www.esrcsocietytoday.ac.uk/ESRCInfoCentre/Images/bill_greenhalgh_tc
  m6-9605.ppt. [Date: 6/16/09]

- Holiday Rentals (2009): Location: Seattle in USA. Available at:
  http://www.holiday-rentals.co.uk/p226819 [Date: 6/22/09]

- Lancashire County Council (2008): The Aerospace Industry in Lancashire.
  Available at:
  http://www.lancashire.gov.uk/office_of_the_chief_executive/lancashireprofile/s
  ectors/aero.asp?sysredir=y [Date: 6/16/09]

- Northwest Aerospace Alliance (2007): The Aerospace Cluster Strategy 2007 to
  2017. Available at:
  http://www.nwda.co.uk/pdf/The%20Aerospace%20Cluster%20Strategy%20200
  7%20Issue%201%20Oct%2007.pdf [Date: 6/16/09]

- North West Development Agency (NWDA) (2004): Aerospace Insider Business
  Guide. Available at:
  http://www.northwestscience.co.uk/uploads/documents/apr_06/northwestscience
  _1144666206_Aerospace2004.pdf. [Date: 6/22/09]